Aircraft

Aircraft

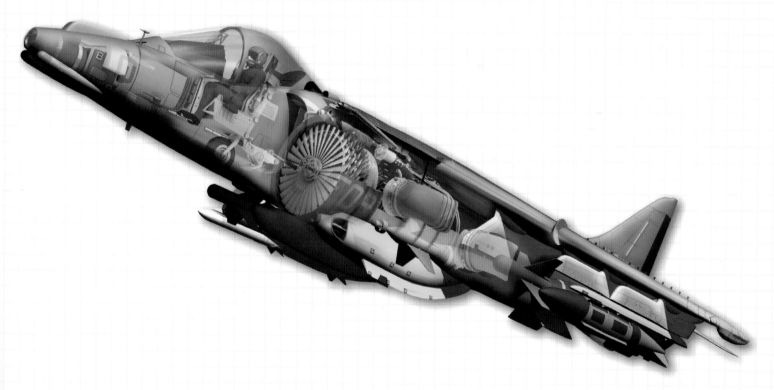

By Steve Parker
Illustrated by Alex Pang

MASON CREST PUBLISHERS INC.
370 Reed Road, Broomall, Pennsylvania 19008
(866)MCP-BOOK (toll free), www.masoncrest.com

First Printing
9 8 7 6 5 4 3 2 1

        Library of Congress Cataloging-in-Publication Data
Parker, Steve, 1952–
  Aircraft / by Steve Parker ; illustrations by Alex Pang.
     p. cm. — (How it works)
  Includes bibliographical references and index.
  ISBN 978-1-4222-1791-7
  Series ISBN (10 titles): 978-1-4222-1790-0
  1. Airplanes—Juvenile literature. I. Pang, Alex, ill. II. Title.
  TL547.P267 2011
  629.133'34—dc22
                        2010033385

Printed in the U.S.A.

First published by Miles Kelly Publishing Ltd
Bardfield Centre, Great Bardfield, Essex, CM7 4SL
© 2009 Miles Kelly Publishing Ltd

Editorial Director: Belinda Gallagher
Art Director: Jo Brewer
Design Concept: Simon Lee
Volume Design: Rocket Design
Cover Designer: Simon Lee
Indexer: Gill Lee
Americanizer: Betty Christiansen
Production Manager: Elizabeth Brunwin
Reprographics: Stephan Davis, Ian Paulyn
Consultants: John and Sue Becklake

Every effort has been made to acknowledge the source and copyright
holder of each picture. The publisher apologizes for any unintentional
errors or omissions.

ACKNOWLEDGMENTS

All panel artworks by Rocket Design
The publishers would like to thank the following
sources for the use of their photographs:
Alamy: 30 Emil Pozar III
Aviation Images: 32 M Wagner
Corbis: 17 Aero Graphics, Inc.; 21 Bettmann;
23 Hulton-Deutsch Collection; 25 Antonio Cotrim/epa;
29 Handout/Reuters
Fotolia: 9 Charles Shapiro; 13 Igor Zhorov
Getty Images: 36 Johnny Green
Rex Features: 6 Jonathan Hordle; 7 Hugh W. Cowin;
15 C.WisHisSoc/Everett; 27
All other photographs are from Miles Kelly Archives

WWW.FACTSFORPROJECTS.COM

Each top right-hand page directs
you to the Internet to help you
find out more. You can log on
to **www.factsforprojects.com**
to find free pictures, additional
information, videos, fun
activities, and further web links.
These are for your own personal
use and should not be copied or
distributed for any commercial
or profit-related purpose.

If you do decide to use the
Internet with your book, here's a
list of what you'll need:
• A PC with Microsoft® Windows®
  XP or later versions, or a
  Macintosh with OS X or later,
  and 512Mb RAM

• A browser such as Microsoft®
  Internet Explorer 8, Firefox 3.X,
  or Safari 4.X
• Connection to the Internet via
  a modem (preferably 56Kbps) or
  a faster Broadband connection
• An account with an Internet
  Service Provider (ISP)
• A sound card for listening to
  sound files

Links won't work?
www.factsforprojects.com is
regularly checked to make sure
the links provide you with lots
of information. Sometimes you
may receive a message saying
that a site is unavailable. If this
happens, just try again later.

Stay safe!
When using the Internet, make
sure you follow these guidelines:
• Ask a parent's or a guardian's
  permission before you log on.
• Never give out your personal
  details, such as your name,
  address, or e-mail.
• If a site asks you to log in or
  register by typing your name or
  e-mail address, speak to your
  parent or guardian first.
• If you do receive an e-mail from
  someone you don't know, tell
  an adult and do not reply to the
  message.
• Never arrange to meet anyone
  you have talked to on the
  Internet.

The publisher is not responsible
for the accuracy or suitability
of the information on any
website other than its own. We
recommend that children are
supervised while on the Internet
and that they do not use Internet
chat rooms.

CONTENTS

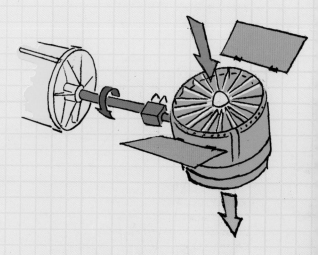

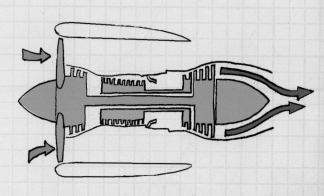

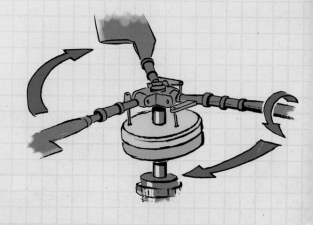

INTRODUCTION

It's an age-old dream to fly like the birds. People once strapped wings made of feathers or cloth to their arms, jumped off cliffs and church towers, and flapped as hard as they could. No one flew, and several plunged to their deaths. Gradually, people realized that human muscles are too weak for the body to fly on its own, even with wings. It needs help from machinery and technology.

Birds inspired early human flyers.

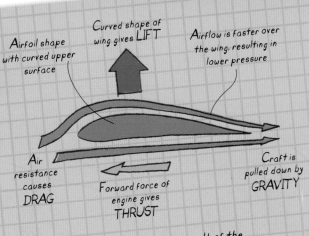

Airfoil shape with curved upper surface

Curved shape of wing gives LIFT

Airflow is faster over the wing, resulting in lower pressure

Air resistance causes DRAG

Forward force of engine gives THRUST

Craft is pulled down by GRAVITY

An aircraft's movements are the result of the balance of four forces: lift, gravity, thrust, and drag.

LIFTOFF

More than 200 years ago, the first people left the ground. They were balloonists, and they stayed airborne for hours, even days. However, they could only go where the wind took them. More than 100 years ago, pioneer pilots built wood-and-cloth craft—the first gliders or sailplanes. They copied the curved airfoil shape of bird wings to give lift. They saw how wings twist, or warp, to give a balance of forces that controls speed and direction. But without power, they couldn't stay airborne for long.

The Wrights' Flyer and other early aircraft were light, delicate, and easily blown around by the wind. Flying was limited to dry, calm days.

PLANES TAKE OFF

In 1903, the Wright Brothers, from Dayton, Ohio, added a small, light, homemade gasoline engine to their carefully tested glider. The engine turned propellers to provide a forward force called thrust. The craft's wings twisted to control its direction. The brothers' spindly machine not only flew, but began a new era of travel, leisure, and danger— the Age of Aircraft.

The aircraft featured in this book are Internet linked.
Visit www.factsforprojects.com to find out more.

ADVANCES IN WAR

During World War I (1914–1918), planes became bigger, stronger, more controllable, and more reliable. After the conflict, more people took up flying and set all kinds of aerial records. World War II (1939–1945) brought more progress, especially a new design of engine to produce the forward force of thrust. It was named after the hot blast of gases from its rear end—the jet.

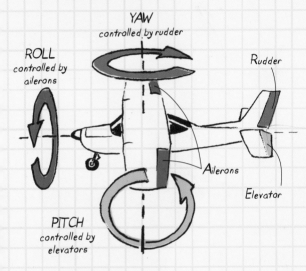

YAW
controlled by rudder

ROLL
controlled by
ailerons

Rudder

Ailerons

Elevator

PITCH
controlled by
elevators

World War II was the first major conflict where aircraft, like the Supermarine Spitfire, played leading roles.

TOWARD TOMORROW

As more peaceful times arrived during the 1950s, air travel really took off. Jumbo-sized passenger jets now carry people around the world for business meetings and exotic vacations. Military planes have become faster and sleeker, with stealth technology to avoid radar and heat-seeking missiles. Some craft take off and land straight up and down, with hovering jets or whirling helicopter rotors. Like everything else, computers get in on the act, as "fly-by-wire" helps pilots and their planes to stay safer.

The Airbus A380 Super Jumbo is the latest and biggest in a long line of passenger aircraft

The world of aircraft never stands still. Aviation is still hardly more than one century old. What will the next 100 years bring?

HOT-AIR BALLOON

Balloons do not truly fly—they float. A hot-air balloon contains air that is heated by a burner. The heat makes the tiny particles (molecules) of air spread out so there are fewer of them in the same space. This makes the air inside the balloon lighter than the air around it, so the balloon rises. The pilot can control the balloon's height by turning up the flame, but its direction depends entirely on the wind.

Eureka!

The first people to make a flight in a hot-air balloon were Pîlatre de Rozier and Francois Laurent, Marquis d'Arlandes. In 1783 in France, they travelled 5.5 miles (8.9 km) across the city of Paris, at a height of about 80 feet (24.4 m), in a balloon made by brothers Joseph and Jacques Montgolfier.

What next?

Scientists have tested a personal helium balloon attached to a backpack with an electric motor and propeller. The idea is that people can fly where they want.

Turning vent

Balloon safaris over the African plains are a good way for tourists to see wildlife.

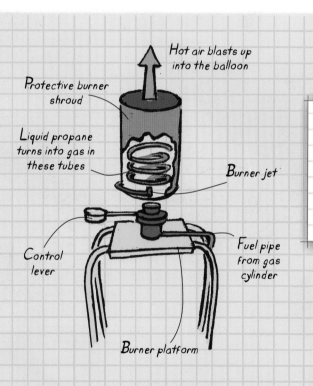

Hot air blasts up into the balloon

Protective burner shroud

Liquid propane turns into gas in these tubes

Burner jet

Control lever

Fuel pipe from gas cylinder

Burner platform

Envelope The balloon's main part is the tough outer casing, the envelope. It's made of long, curved strips, called gores, and nontearable fabric such as nylon. The part near the burner—the throat—is made of heatproof material such as Nomex.

In 2007, David Hempleman-Adams soared to a record height for a hot-air balloon—6 mi.

✳ How do BURNERS work?

Most balloons have a burner that uses propane gas as fuel. The propane is compressed (squeezed) into a liquid inside a metal cylinder. It flows to the burner along a coiled tube, where it's heated by the flame nearby. Then it squirts out of a jet (hole) as a gas, where it mixes with air and catches fire to make the flame.

Load ropes Strong cords sewn into the seams between the panels extend down to the skirt. They spread the weight of the basket around the envelope, like a giant net.

Old airships used hydrogen, which is a very light gas, but this can catch fire. The Hindenberg went up in flames in 1937, killing 36 people and ending the airship era.

For a timeline and history of ballooning, visit
www.factsforprojects.com and click the web link.

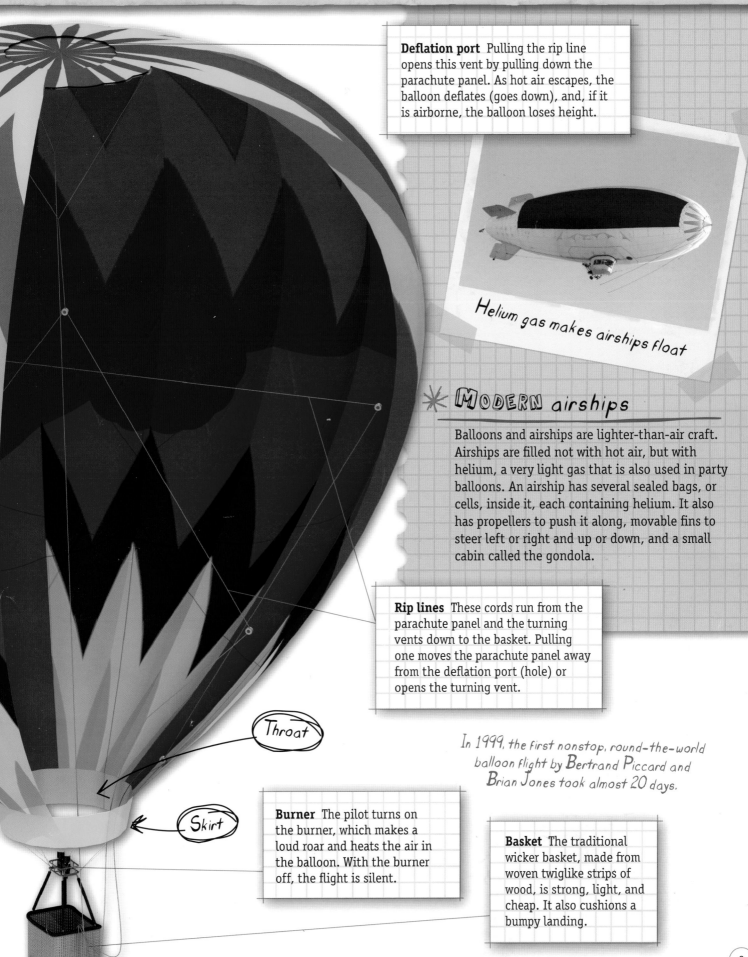

Deflation port Pulling the rip line opens this vent by pulling down the parachute panel. As hot air escapes, the balloon deflates (goes down), and, if it is airborne, the balloon loses height.

Helium gas makes airships float

✳ MODERN airships

Balloons and airships are lighter-than-air craft. Airships are filled not with hot air, but with helium, a very light gas that is also used in party balloons. An airship has several sealed bags, or cells, inside it, each containing helium. It also has propellers to push it along, movable fins to steer left or right and up or down, and a small cabin called the gondola.

Rip lines These cords run from the parachute panel and the turning vents down to the basket. Pulling one moves the parachute panel away from the deflation port (hole) or opens the turning vent.

Throat

In 1999, the first nonstop, round-the-world balloon flight by Bertrand Piccard and Brian Jones took almost 20 days.

Skirt

Burner The pilot turns on the burner, which makes a loud roar and heats the air in the balloon. With the burner off, the flight is silent.

Basket The traditional wicker basket, made from woven twiglike strips of wood, is strong, light, and cheap. It also cushions a bumpy landing.

HANG GLIDER

A glider is an aircraft without an engine. A hang glider is a fabric wing stretched over a rigid frame from which the pilot hangs by straps. It can be steered left and right and go fast or slow. However, it cannot climb unaided (unless carried by rising air), so it gradually descends through the air around it.

Eureka!

Hang gliding was pioneered more than a century ago by Otto Lilienthal near Berlin, Germany. He built his own hill and 20 gliders of his own design and made more than 2,500 flights. Sadly, in 1896, he crash-landed and died.

Leading edge tubes Strong tubes along the wing's front give it the correct shape. Like all parts of the frame, they are made of light but strong aluminum metal or carbon-fiber composite.

Bracing wires Thin, strong metal wires join many sections of the hang glider frame. They strengthen it and help it keep its shape even in high winds and during fast turns.

Nose

How do HANG GLIDER controls work?

The pilot lies in a harness inside a bag, or cocoon, with body weight directly below the main wing for straight, level flight. Moving the control bar shifts the body weight from this central position and tips the wing at an angle.

A-frame

Pushing the control bar forward moves the body weight back and makes the craft rise

Swinging the control bar right shifts body weight to the left, and the craft turns left

Swinging the control bar left shifts body weight to the right, and the craft turns right

Pulling the control bar moves body weight forward and the craft tilts nose-down to descend faster

A-frame This metal frame is fixed by a firm, rigid joint at its top end to the hang glider's main lengthwise pole, the keel.

Control bar Part of the A-frame, the pilot holds the control bar by the nonslip grips and pushes or pulls it to make the hang glider fly in different directions.

In 2002 in Texas, Michael Barber flew a hang glider more than 435 mi (700.1 km).

What next?

A powered hang glider has a small, lightweight motorcycle-type engine just behind the pilot, which turns a small propeller. It's the closest machine that most of us can get to a personal aircraft.

If you're interested in learning to hang glide, visit www.factsforprojects.com and click the web link.

Crossbar

Plastic battens Long, thin strips called battens slide into pockets in the wing. These keep the wing straight and rigid in the right places, so it slips easily through the air without flapping.

Wing (sail) The shape of a hang glider is based on a design called the Rogallo wing. It's also made of very strong tear-proof material such as nylon or Kevlar.

The fastest hang gliders can exceed speeds of 85 mph (136.8 km/h).

Keel

In the best conditions with lots of rising air, hang glider pilots can reach heights of more than 16,000 ft (4,876.8 m).

✳ FLY like a bird!

Gliders cannot climb under their own power—they have none. They slowly descend through the air. However, if the pilot can find air that's rising—like wind blowing up a hill or hot air rising from sun-warmed rocks—this will carry the glider higher. After swooping down, the pilot can do the same again and in favorable conditions stay up in the air for hours.

Air rising over mountains lifts a hang glider higher

SAILPLANE

L ike any type of glider, the sailplane cannot rise through the air under its own power since it has no motor or engine. However, the modern high-performance sailplane is so light and streamlined that it loses height very, very slowly as it glides. If it finds rising air—such as wind blowing up a hill—it soon climbs up near the clouds.

Eureka!

Most sailplanes have a skin of light, smooth, strong fiberglass (GRP, glass-reinforced plastic resin). This was first used in the German Akaflieg FS-24 glider in 1957.

What next?

The newest gliders use carbon, aramid, and polyethylene fibers for construction and have detachable upturned wing ends or "twisted winglets" as found on passenger jets.

Retractable undercarriage
The single wheel folds up into the body after takeoff and until landing to reduce air resistance, or drag.

Construction
Sailplanes make use of the strongest, lightest materials such as carbon-fiber composites and alloys (mixtures of metals).

Cockpit The pilot has fewer controls and displays than in a powered plane since there are no levers or dials for engine speed, fuel level, and similar readings.

Tow point

Fiberglass skin

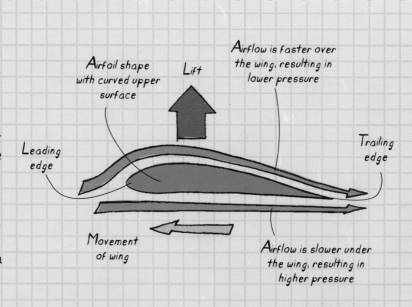

✳ How do WINGS work?

An aircraft's wing is not a flat sheet. It is curved when seen from the edge, a shape called an airfoil. The curve is greater on the upper surface than the lower. Air flowing past the wing must go faster over the top than beneath, and faster air means less air pressure. So the lower air pressure above the wing sucks it upward with a force called lift.

Airfoil shape with curved upper surface

Lift

Airflow is faster over the wing, resulting in lower pressure

Leading edge

Trailing edge

Movement of wing

Airflow is slower under the wing, resulting in higher pressure

Visit www.factsforprojects.com and click the web link to find lots of paper gliders to make and design.

During World War II (1939–1945), huge gliders landed troops and vehicles in emy territory hout a sound.

Tailplane As a sailplane has just a single wheel for its undercarriage, the whole craft may rock from side to side on takeoff and landing. The high-set tailplane is less likely to clip the ground compared to a low-set one.

Streamlined shape

Fin ribs

Rudder

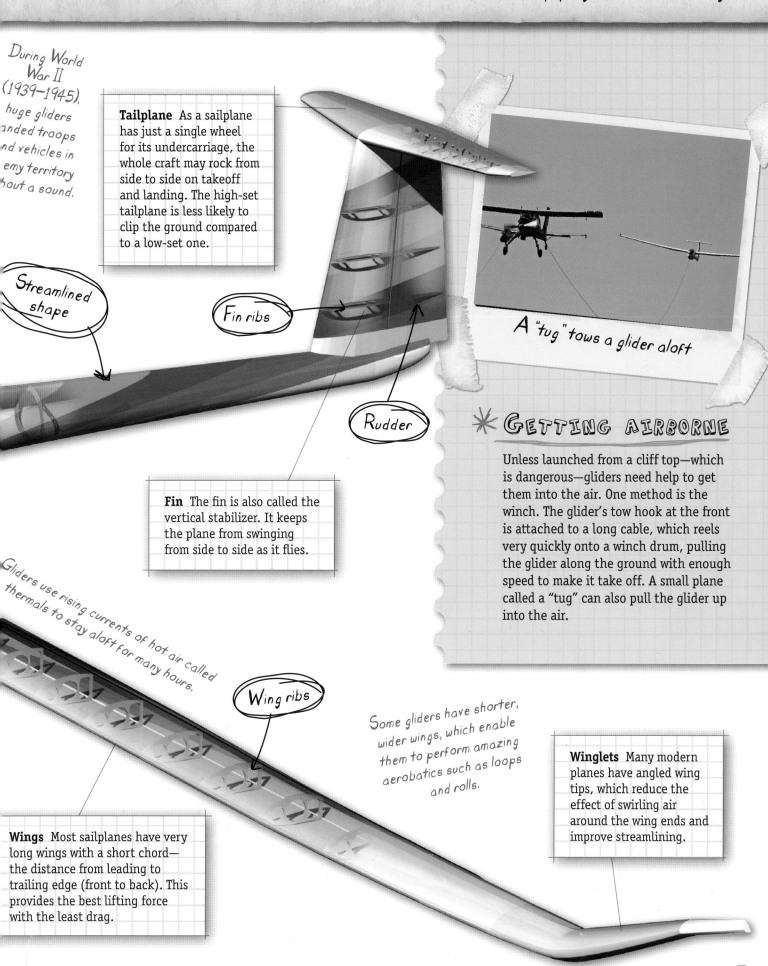

A "tug" tows a glider aloft

Fin The fin is also called the vertical stabilizer. It keeps the plane from swinging from side to side as it flies.

✳ GETTING AIRBORNE

Unless launched from a cliff top—which is dangerous—gliders need help to get them into the air. One method is the winch. The glider's tow hook at the front is attached to a long cable, which reels very quickly onto a winch drum, pulling the glider along the ground with enough speed to make it take off. A small plane called a "tug" can also pull the glider up into the air.

Gliders use rising currents of hot air called thermals to stay aloft for many hours.

Wing ribs

Some gliders have shorter, wider wings, which enable them to perform amazing aerobatics such as loops and rolls.

Winglets Many modern planes have angled wing tips, which reduce the effect of swirling air around the wing ends and improve streamlining.

Wings Most sailplanes have very long wings with a short chord— the distance from leading to trailing edge (front to back). This provides the best lifting force with the least drag.

WRIGHT FLYER

The time: 10:35 a.m., Thursday, December 17, 1903. The place: a windy beach near Kitty Hawk, North Carolina. The event: the first-ever flight in a controlled, powered aircraft, the *Flyer*, built by brothers Wilbur and Orville Wright. Watched by just a handful of helpers, it lasted 12 seconds—but it would change the world.

Eureka!

The Wright brothers spent many hours watching how birds twist their wings for flight control. This gave them the wing-warping idea to control the *Flyer*.

What next?

After Orville's first flight, the *Flyer* made three more trips that day, with the brothers taking turns flying. Wilbur's last flight was the longest—850 feet (259.1 m) in 59 seconds.

Muslin covering

The *Flyer* covered 121 ft (36.9 m) on its first trip—half the length of an Airbus A380 jumbo jet.

Elevator lever
The pilot's left hand worked a lever to tilt the two front elevators and control the *Flyer's* height.

Canard layout
Flyer had elevators at the front, which is known as a canard design.

Elevators

Hip cradle

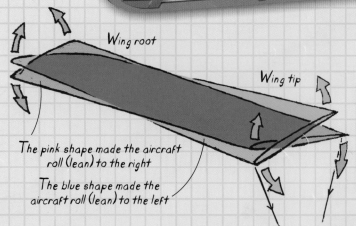

Wing root

Wing tip

The pink shape made the aircraft roll (lean) to the right

The blue shape made the aircraft roll (lean) to the left

Wires pulled the wing to twist its shape

Skids The base of the main airframe worked like skis, or skids, to slide along the sand when landing.

✳ How did WING WARPING work?

Flyer's flexible wings could be twisted, or warped, along their length. This gave the wing on one side more lifting force than the other, raising it higher as the other wing dipped. It made the plane lean or roll to the side. Modern planes have flap-like ailerons at the rear outer edges of their wings for the same purpose (see page 16).

To find out how the Wright Flyer was recreated for its 100th birthday, visit www.factsforprojects.com and click the web link.

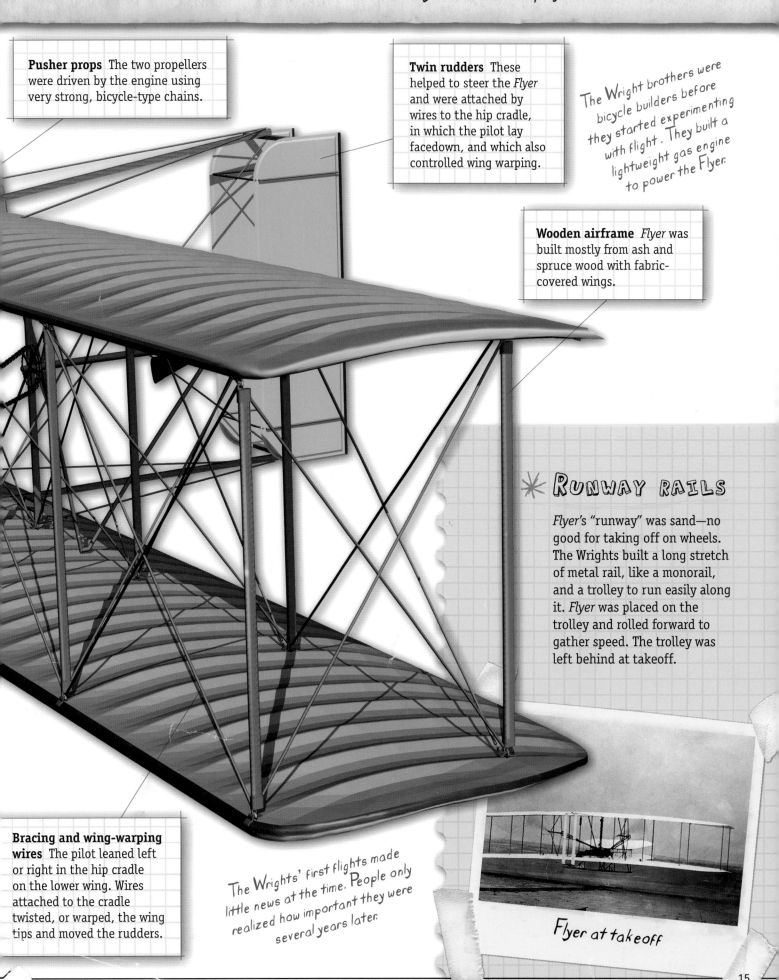

Pusher props The two propellers were driven by the engine using very strong, bicycle-type chains.

Twin rudders These helped to steer the *Flyer* and were attached by wires to the hip cradle, in which the pilot lay facedown, and which also controlled wing warping.

The Wright brothers were bicycle builders before they started experimenting with flight. They built a lightweight gas engine to power the Flyer.

Wooden airframe *Flyer* was built mostly from ash and spruce wood with fabric-covered wings.

✳ RUNWAY RAILS

Flyer's "runway" was sand—no good for taking off on wheels. The Wrights built a long stretch of metal rail, like a monorail, and a trolley to run easily along it. *Flyer* was placed on the trolley and rolled forward to gather speed. The trolley was left behind at takeoff.

Bracing and wing-warping wires The pilot leaned left or right in the hip cradle on the lower wing. Wires attached to the cradle twisted, or warped, the wing tips and moved the rudders.

The Wrights' first flights made little news at the time. People only realized how important they were several years later.

Flyer at takeoff

15

CESSNA 172 SKYHAWK

Light aircraft are like flying versions of family cars. The most common kinds have four seats and a single car-type engine that spins a propeller. These planes cannot fly very fast or do amazing stunts. However, they are strong, reliable, easy to fly, and simple to service. They are also light—about half the weight of a family car.

Eureka!

The first Cessna 172s were built more than 50 years ago, in 1955. From the start, the makers believed they had a superior design—and they were right.

What next?

A future version of the Cessna 172 will be powered by a turbo diesel engine, similar to the engines in some fast cars.

Engine The plane is powered mainly by Lycoming or Continental engines with four or six cylinders. The engine works in the same way as a car engine. Fuel and air burn inside the cylinders and push the pistons along. However, the cylinders are not in a line, as in a car. They lie flat, with one half facing one way and the other half facing the other way. This is called a horizontally opposed layout.

Control stick This is linked to the ailerons and elevators by long metal cables. Push it forward and the plane dives; pull it back and the plane climbs.

Spinner

Rudder pedals Pushing the left pedal causes the rudder at the rear to swing to the left, making the plane steer to the left.

Yaw
controlled by rudder

Roll
controlled by
ailerons

Rudder

Ailerons

Elevator

Pitch
controlled by
elevators

✳ How do CONTROL SURFACES work?

Control surfaces are movable parts on the wings and tail. As a control surface moves, air rushing past pushes against it and moves that part of the plane. This changes the plane's pitch (up and down), yaw (left or right), and roll (lean to the side).

More Cessna 172s have been built than any other aircraft—more than 43,00

To watch amazing videos of a Cessna in flight, visit
www.factsforprojects.com and click the web link.

More people have learned to fly in the
Cessna 172 than in any other plane.

Elevators These control
surfaces are hinged onto
the tailplane, the small wing
at the rear. They are moved
by long cables attached to
the control stick. They both
move up or down together.

Rudder This control surface
is worked by the pilot's pedals
and makes the plane steer left
or right. It is hinged onto the
upright "tail," which is known
as the fin, and helps the plane
to fly straight.

Tailplane

The name
"Skyhawk" was
first used for the
Cessna 172 in
1961.

Elevator cables

Rudder cables

Control cable
pulleys

The 172's normal cruising
speed is 137 mph (220.5
km/h) but versions with
a souped-up engine can
reach a speed of almost
155 mph (249.4 km/h).

Wheel fairing

✳ How do TAILLESS aircraft fly?

Delta-wing or flying-wing planes
have no tailplane at the rear for the
elevators. Instead, each main wing has
a combined elevator-aileron control
surface, the elevon.

Elevon

Ailerons These control
surfaces are at the end
of each wing, on the rear
or trailing edge. They are
worked by cables from the
control stick. When the
aileron on one side tilts up,
the other tilts down.

F-117 Nighthawk fighter,
a delta-wing aircraft

SOPWITH CAMEL

One of the best fighter planes of World War I (1914–1918), the Camel was speedy and agile in the air. As they twisted and turned in deadly aerial battles called "dogfights," Camels from the air forces of Britain and its allies (countries on the same side of the war) shot down almost 1,300 enemy planes—more than any other allied aircraft.

Twin machine guns The pilot aimed the Vickers machine guns accurately by looking along their barrels and pointing the plane straight at the target.

Biplane design A biplane has two sets of main wings, one above the other. This gives lots of lifting force for a fairly small wingspan (tip-to-tip length).

Struts and braces Wooden pole-like struts held apart the two sets of wings. Taut bracing wires made the whole structure very strong yet lightweight.

Rotary engine The cylinders were arranged like the spokes of a wheel, as in a radial engine (see page 20). They moved around the central shaft rather than staying still.

Cock[...]

Engine cowl

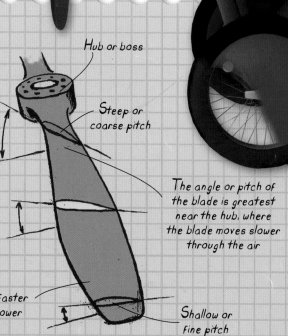

Hub or boss

Steep or coarse pitch

The angle or pitch of the blade is greatest near the hub, where the blade moves slower through the air

Shallow or fine pitch

The engine, fuel, machine guns, and pilot were close together at the front. This made the Camel tricky to learn to fly but a suprem[e] dogfighter fo[r] a fighter "ac[e]"

Fabric-covered, spoked wheels

✳ How do PROPELLERS work?

A propeller works partly by pushing air backward, like a spinning electric fan with flat, angled blades. Propeller blades are not flat. They have an airfoil shape (see the blue, yellow, and green cutaway areas) more curved on the front surface than the rear, like an airplane wing. This produces lower air pressure in front of the blade than behind. So the blade "sucks" itself forward as well as pushing air backward.

The blade tip moves much faster through the air, so a shallower angle works best

Eureka!

To prevent their own bullets from shooting off their propellers, early fighting planes had metal wedge shapes called deflectors fixed to the backs of the propeller blades. The bullets bounced off these at an angle rather than smashing into the wooden blade.

What next?

The 2F.1 version of the Camel could take off from boats and carried out the first ever ship-launched air raids in 1918.

Find out more about vintage aircraft by visiting www.factsforprojects.com and clicking the web link.

The Camel took its name from the humplike cover over the machine guns.

Roundel This was the symbol of Britain's Royal Flying Corps, which became the Royal Air Force in 1918. It identified the plane so allies would not attack it.

Fin and tailplane The fixed surfaces of these parts had bracing wires to keep them rigid and in position.

Rudder

Squadron markings

U—F2227

A Camel was tested as a mid-air launched fighter by attaching it under a huge airship and simply dropping it.

Wooden airframe

Nearly 5,500 Camels were built, among the most of any World War I plane.

☀ Don't shoot the PROP!

Very early warplanes had a handheld gun, then a hand-operated gun on a metal arm with a swivel joint, aimed by the pilot or gunner. Sometimes in the heat of combat, these flyers shot off their own propellers by mistake! The Camel had a synchronization mechanism in which the spinning propeller shaft had a bulge or cam that pushed a rod each time it turned. The rod pressed against the trigger and prevented the machine gun from firing for a split second each time a blade was in the way.

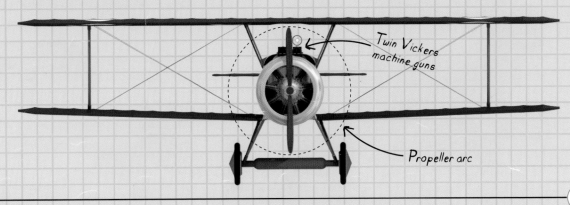

Twin Vickers machine guns

Propeller arc

RYAN NYP SPIRIT OF ST. LOUIS

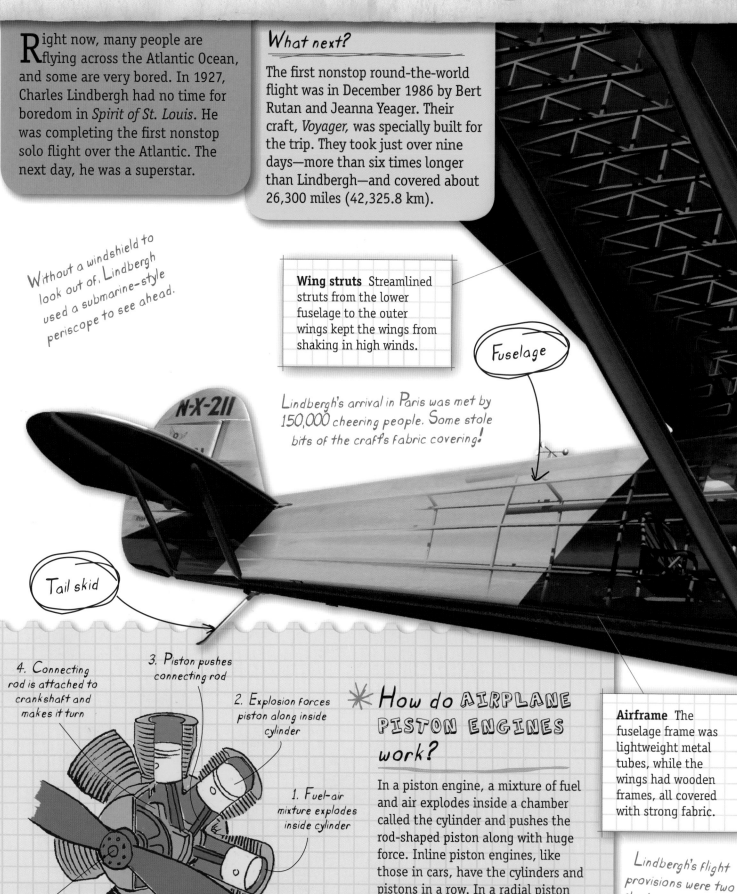

Right now, many people are flying across the Atlantic Ocean, and some are very bored. In 1927, Charles Lindbergh had no time for boredom in *Spirit of St. Louis*. He was completing the first nonstop solo flight over the Atlantic. The next day, he was a superstar.

What next?

The first nonstop round-the-world flight was in December 1986 by Bert Rutan and Jeanna Yeager. Their craft, *Voyager*, was specially built for the trip. They took just over nine days—more than six times longer than Lindbergh—and covered about 26,300 miles (42,325.8 km).

Without a windshield to look out of, Lindbergh used a submarine-style periscope to see ahead.

Wing struts Streamlined struts from the lower fuselage to the outer wings kept the wings from shaking in high winds.

Fuselage

N-X-211

Lindbergh's arrival in Paris was met by 150,000 cheering people. Some stole bits of the craft's fabric covering!

Tail skid

4. Connecting rod is attached to crankshaft and makes it turn

3. Piston pushes connecting rod

2. Explosion forces piston along inside cylinder

1. Fuel-air mixture explodes inside cylinder

5. Propeller is mounted on crankshaft

Piston

✳ How do AIRPLANE PISTON ENGINES work?

In a piston engine, a mixture of fuel and air explodes inside a chamber called the cylinder and pushes the rod-shaped piston along with huge force. Inline piston engines, like those in cars, have the cylinders and pistons in a row. In a radial piston engine, the cylinders are arranged in a circle, like the spokes of a wheel.

Airframe The fuselage frame was lightweight metal tubes, while the wings had wooden frames, all covered with strong fabric.

Lindbergh's flight provisions were two flasks of water and four ham sandwiches.

Discover everything you need to know about Lindbergh and his epic flight at www.factsforprojects.com

Monoplane design The *Spirit* was a "shoulder monoplane" with one pair of main wings fixed to the upper sides of the fuselage. Its wingspan (distance from tip to tip) was 46 feet (14m).

The *Spirit's* letters "NYP" stood for New York–Paris.

No windshield!

Cockpit The *Spirit* was a modified M2 mail plane with a cockpit so small that Lindbergh could not stretch his legs. Dials and instruments were on the rear of the massive 450-gallon (1,703.4 L) fuel tank.

Cockpit

Wright J-5C Whirlwind This powerful, reliable, well-tested radial engine had nine cylinders and produced 223 horsepower, more than most family cars today.

Spirit of St. Louis

* LUCKY LINDY!

Lindbergh won the Orteig Prize of $25,000 for the first solo nonstop flight between New York and Paris. Six pilots had already died trying when he took off from Roosevelt Airfield at 7:52 a.m. on May 20. He landed 33 hours 29 minutes later at Le Bourget Aerodrome, Paris. It was dark, at 10:22 p.m. on the evening of the next day.

The journey of Lucky Lindy, the "Lone Eagle"

Lindbergh almost plunged in the water when the *Spirit*, weighed down with sleet and ice, dove to only 10 ft (3 m) above the waves.

Streamlined wheels

SUPERMARINE SPITFIRE

One of the world's most famous aircraft, the Spitfire fighter plane first flew in 1936 and was still being produced nine years later. It played a major part during World War II (1939–1945), especially during the Battle of Britain (1940), being just about the fastest and most agile warplane of the time. Spitfires stayed in service until the 1950s.

Eureka!

The fastest Spitfires were able to chase jet-powered V1 flying bombs, fly alongside them, and wiggle their wing tips to flip the bombs off course.

Over 20,000 Spitfires were built—more than any similar warplane.

Engine Early Spitfires had a Rolls Royce Merlin, as used on other aircraft of the time such as the Lancaster bomber. Later Spitfires had Rolls Royce's more powerful Griffon engine.

Fuel tank

✳ How do adjustable PITCH PROPS work?

The exact size, shape, and pitch (angle) of propeller blades are carefully designed to give the best forward force or thrust. However, as a prop turns faster for higher-speed flight, a different pitch works better than for slow flight. The adjustable pitch prop automatically changes the pitch of its blades to give the most possible thrust at different spinning speeds.

Coarse (steep) pitch for high speed flight

Propeller blade twists along this axis (line)

Streamlined spinner

Fine (shallow) pitch for low speed flight

Armament Early Spitfires had Browning machine guns in each wing. Later ones were also fitted with the more powerful Hispano cannons.

Propeller Different versions of the Spitfire had propellers with different numbers of blades from two to six. These became larger with more powerful engines.

The fastest wartime Spitfire had a top speed of about 450 mph (724.2 km/h). Built for spying and photographic missions, it was painted pink to help it blend in with sunrise and sunset.

What next?

Around the world about 50 Spitfires are still flying today. A few have a second seat so a passenger can enjoy the ride.

Watch a video to find out exactly what it takes to become a Spitfire pilot by visiting www.factsforprojects.com and clicking on the web link.

Several names were considered for the Spitfire and it was almost called the Supermarine Shrew!

Stressed skin The outer covering of lightweight aluminum-based metal sheet helped to cope with stresses and strains.

Cockpit

ROLLS ROYCE MERLIN

The Spitfire's Merlin engine was 27 liters (a typical family car is 2 liters). It had 12 cylinders in two rows of six at an angle to each other called a V12. Its design and fuel were continually improved over the years. In early Spitfires, it produced about 1,000 horsepower. Ten years later this power had doubled.

Merlin-engined Spitfires on the production line

Retractable landing gear The main wheels folded up into the wings after takeoff to reduce drag, or air resistance, so the plane could fly faster and farther.

Curved wing tips

Wing shape The wings had a distinctive elliptical or oval shape when seen from above or below (what is called planform view).

Spitfires were famous in the Battle of Britain. However, their companion fighters, Hawker Hurricanes, shot down more enemies.

SHORT SUNDERLAND S.25

Before big airports with hard runways, huge planes had trouble landing on airfields. "Flying boat" seaplanes could use any sizable stretch of water, such as a river, lake, or ocean. From World War II (1939–1945) until the 1960s, long-range Sunderlands patrolled the oceans to spot dangers such as enemy submarines, and some also carried passengers in great luxury.

Eureka!

In 1914, the world's first regular airline service began in the U.S.A. The St. Petersburg–Tampa Airboat Line used Benoist XIV flying boats. Each trip took 10 minutes and carried up to three passengers.

What next?

Many inventors have built flying cars that can take off from an ordinary straight road. The problem is that learning to fly takes at least 50 times longer and is more than 20 times more expensive than learning to drive.

Top turret

Gun turrets The gun turrets had Browning .303 machine guns and could swivel around under electrical power to aim at the enemy.

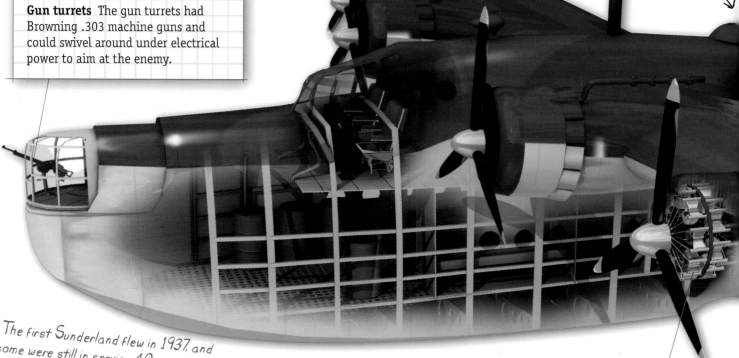

The first Sunderland flew in 1937, and some were still in service 40 years later.

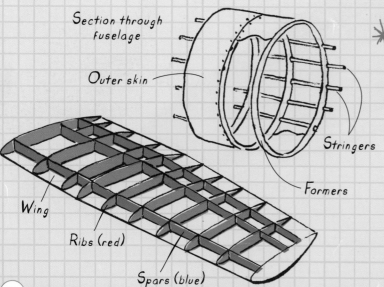

Section through fuselage

Outer skin

Stringers

Formers

Wing

Ribs (red)

Spars (blue)

✴ How does an AIRFRAME work?

The airframe is the strong "skeleton" of an aircraft. The fuselage has hoop-shaped formers and long stringers, or longerons. The wings have curved ribs from front to back and rigid spars along their length. Both are usually covered in a thin skin of the very light metal aluminum.

Engines The most powerful Sunderlands had four Pratt & Whitney R-1830-90B Twin Wasp engines, each with 14 cylinders.

The Sunderland's fuselage was shaped like a fast boat so it could take off and land in the waves.

Discover more about the history of the Sunderland by visiting www.factsforprojects.com and clicking in the web link.

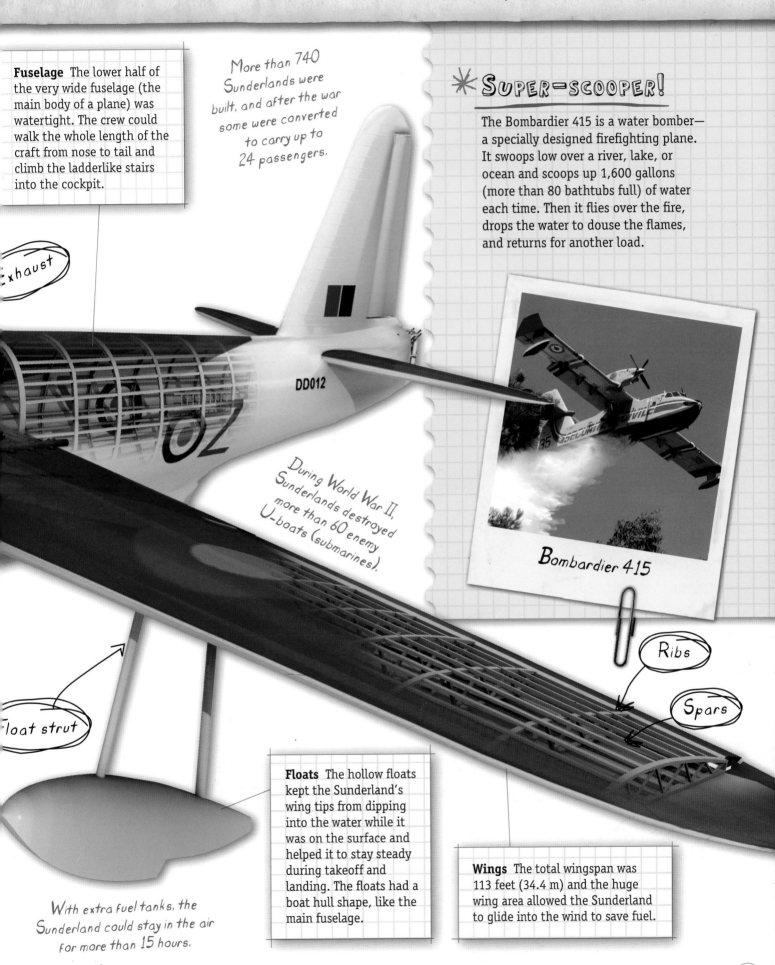

Fuselage The lower half of the very wide fuselage (the main body of a plane) was watertight. The crew could walk the whole length of the craft from nose to tail and climb the ladderlike stairs into the cockpit.

More than 740 Sunderlands were built, and after the war some were converted to carry up to 24 passengers.

Exhaust

DD012

During World War II, Sunderlands destroyed more than 60 enemy U-boats (submarines).

✳ SUPER-SCOOPER!

The Bombardier 415 is a water bomber— a specially designed firefighting plane. It swoops low over a river, lake, or ocean and scoops up 1,600 gallons (more than 80 bathtubs full) of water each time. Then it flies over the fire, drops the water to douse the flames, and returns for another load.

Bombardier 415

Ribs

Spars

Float strut

With extra fuel tanks, the Sunderland could stay in the air for more than 15 hours.

Floats The hollow floats kept the Sunderland's wing tips from dipping into the water while it was on the surface and helped it to stay steady during takeoff and landing. The floats had a boat hull shape, like the main fuselage.

Wings The total wingspan was 113 feet (34.4 m) and the huge wing area allowed the Sunderland to glide into the wind to save fuel.

HARRIER JUMP JET

The Harrier is the world's most successful VTOL jet aircraft. VTOL stands for Vertical Takeoff and Landing. It means the Harrier can rise straight up on takeoff, hover in midair, then come straight down to land. The first Harriers flew in the late 1960s and have been through four main versions and many small improvements. Dozens still serve in several air forces around the world.

Eureka!

One of the VTOL test craft that led to the Harrier involved two jet engines fitted into a metal frame. It hovered briefly in 1953 and was called the "Flying Bedstead."

Pilot Harrier pilots say that the craft flies like an ordinary plane at fast speeds, but more like a helicopter at slow speeds and when hovering or moving up and down.

Cockpit canopy

VTOL uses so much fuel that Harriers are usually STOL, Short Takeoff and Landing.

Midair refueling nozzl

Nose radar This sends out powerful radio wave blips and detects their echoes bouncing off objects ahead.

How does the HARRIER hover?

The Harrier's single Rolls Royce Pegasus jet engine sends out a continuous blast of hot gases, but these do not come out of a hole at the plane's rear, as usual. They rush out of four underwing nozzles, which can swivel through a right angle (90°). The nozzles point straight down for vertical flight and then slowly turn or swivel to aim straight backward for full speed ahead.

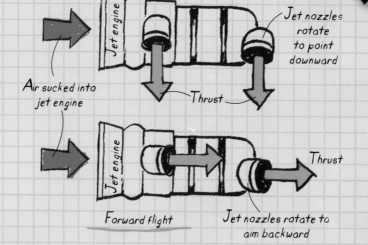

Hovering or vertical flight

Jet engine

Air sucked into jet engine

Thrust

Jet nozzles rotate to point downward

Jet engine

Forward flight

Thrust

Jet nozzles rotate to aim backward

Air intake Air for the jet engine is sucked in through two intakes, one on each side of the fuselage.

What next?

In September 2008, Swiss aviator Yves Rossy flew across the English Channel using a backpack powered by four jet engines. Rossy reached a speed of about 190 miles an hour and completed the 21-mile (33.8-km) journey in less than 10 minutes.

Harriers have starred in several big movies, including James Bond's The Living Daylights (1987) and True Lies (1994) with Arnold Schwarzenegger.

To watch the Harrier at work, visit www.factsforprojects.com and click the web link.

✳ FAST EXIT

Many military aircraft have ejector seats. In an emergency, the pilot pulls a lever that makes the cockpit canopy (cover) blast away and then sets off a small rocket under the seat. The rocket fires the seat with the pilot upward to get clear of the plane—especially the tail, which comes up fast behind. The pilot can then detach from the seat and parachute down to safety.

A pilot makes an emergency exit using his ejector seat

Missile

Underwing pylon mountings

Rear-facing radar
A radar dish in the tail detects planes, ships, and other objects up to several hundred miles behind the Harrier.

Wing tip thruster

Jet thrust nozzles The blast of air from the jet engine comes out through four swiveling nozzles. There are two on either side, one in front of the other.

The Harrier can usually hover for just 90 seconds. Then its jet engine starts to overheat because it is not cooled by moving through the air.

Underwing pods Clip-on pods can carry various weapons such as rockets and missiles, or extra fuel tanks to give a greater range (distance).

LOCKHEED C-130 HERCULES

Wars don't always break out next to large airports with long, smooth runways. Transport planes such as the Hercules must be strong and tough, able to lift great loads after a short takeoff from rough air strips or even from farm fields and sandy deserts. The Hercules first flew in 1954, and hundreds are still in action in more than 60 countries.

Eureka!

After landing, the Russian An-124 Condor heavy transport "kneels" by shortening its nose wheel undercarriage. This lowers the nose to the ground, then the whole nose door swings opens for easy loading.

One Hercules was equipped with rocket slanting downward and missile engines facing backward so it could take off an land almost like a helicopter!

Propellers Early Hercules models had three or four blades for each propeller. The latest version, the C-130J Super Hercules, has six blades on each prop.

Flight deck The standard flight crew of five is a captain (main pilot), a copilot, a navigator, a flight engineer who looks after the engines and mechanical systems, and a loadmaster in charge of the cargo.

More than 2,200 Hercules have been made for 67 countries worldwide.

Forward-looking radar

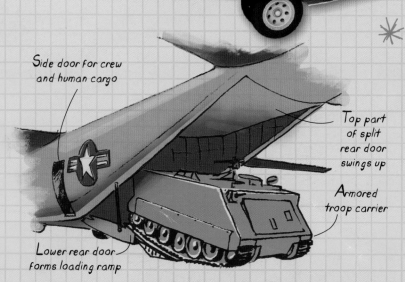

Side door for crew and human cargo

Top part of split rear door swings up

Armored troop carrier

Lower rear door forms loading ramp

✳ LOADING UP

Some versions of the Hercules can carry more than 20 tons (18.1 metric tons) of cargo. Military vehicles such as jeeps and armored cars, or more than 100 combat-ready soldiers, load themselves up the rear ramp, which lowers down to the ground. Or rollers can be fitted onto the floor to slide tray-like pallets of freight inside the fuselage.

Undercarriage bay

The Hercules' many roles include midair refueling tanker, flying hospital, survey plane, airborne weather station, and search and rescue.

What next?

The problem of landing a Hercules in cold, snowy places was solved by fitting it with skis in addition to ordinary wheels.

Watch an amazing video of a Hercules in action by visiting www.factsforprojects.com and clicking the web link.

UF 8668

The Hercules is one of only four planes still in production more than 50 years after it was first introduced.

High-set tailplane The tailplane is set high up at the end of the fuselage to allow for the loading ramp, which forms the rear, sloping part of the fuselage floor.

Engines Four turboprop engines power the Hercules. These have turbine blades inside like a jet engine (see page 32) rather than pistons in cylinders.

Cargo bay The large wings sit on top of the fuselage so that they do not obstruct the flat floor of the cargo area. This is up to 56 feet (17 m) long in "stretched" versions.

Desert color scheme

MULTI-WHEELS

The Hercules' many huge wheels with soft tires spread its fully loaded weight of more than 75 tons (68 metric tons). This means it can roll across soft ground such as a grass air strip. Hercules planes carry loads varying from military vehicles and weapons to emergency provisions for disaster areas.

Extra wheels spread the Hercules' weight

NORTHROP B-2 SPIRIT BOMBER

No other aircraft looks like the B-2 Spirit, a long-distance bomber and spyplane. It has no tailplane, no fin, and no real fuselage, either. Its flying wing shape is specially designed to be extremely difficult to detect as it sneaks up for a surprise attack. Its first flight was in 1989, and the B-2 finally went into service with the U.S. Air Force in 1996.

Eureka!

During World War I (1914–1918), inventors tried to make planes invisible with a covering of clear cellophane instead of cloth. However, the sunlight glinted off the shiny surface.

What next?

Several big aircraft makers regularly test stealth plane shapes such as the United States' Boeing Bird of Prey, Russia's Sukhoi PAK-FA and China's J-XX.

Engine outlets The hot, noisy blast of gases from the four jet engines goes through ducts or openings in the wing, so it shows up less on the enemy's heat-sensing equipment.

Only 21 B-2s have been made at a cost of some $2 billion each.

A fighter plane fires decoy countermeasure flares

☀ HOT STUFF!

Heat-seeking missiles have infrared (heat) sensors that detect and lock onto a strong heat source, such as the gases blasting from an enemy jet or rocket. To counteract this, an aircraft may fire decoy countermeasure flares to attract the missile away from it.

Angled wing tips

Control surfaces The B-2 has elevators and ailerons combined as elevons (see page 17). Each elevon splits into two flaps that open at the same time to work as an air brake, or deceleron.

Learn all about stealth technology by visiting www.factsforprojects.com and clicking the web link.

Ultra-smooth surface The curved surfaces have no projections such as bomb pods or extra fuel tanks. This helps with streamlining and stealth.

The B-2 has a top speed of just 600 mph (965.6 km/h) (much slower than the speed of sound). However, it can fly 6,000 mi (9,656 km) without having to refuel.

Air intakes The intakes for the engines are hidden within the wing's thickness.

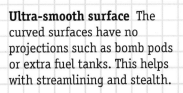

USAF 21001

Cockpit

✳ How does STEALTH work?

A stealth plane such as the B-2 is hard to find by sight, sound, infrared (heat) sensors, or the radio waves of radar. The smooth curves on the top and underside and the "double W" rear edge break up the incoming radar signals into lots of weak beams. These reflect in many directions, rather than as a strong beam that goes back toward the radar equipment.

Flying wing shape With a very thin, low, wide shape and no tail or fuselage bulge, the B-2 is difficult to see from a distance. Its wings measure 171 feet (52.1 m) across, yet its length is just 69 feet (21 m).

Sawtooth shape scatters radar waves in many directions so they are too weak to detect

Regular shape reflects radar waves as a tight beam, which can be detected

Special paint The paint is designed to soak up or absorb the radio waves of radar, rather than bounce them back for detection. However, it is affected by too much heat or cold, so B-2s are kept in air-conditioned hangars.

Leading edge radar-absorbing tape

With midair refueling, B-2s have flown missions of up to 50 hours. Luckily for the two crew, there are on-board hot meals and a flush toilet.

Straight leading and trailing edges

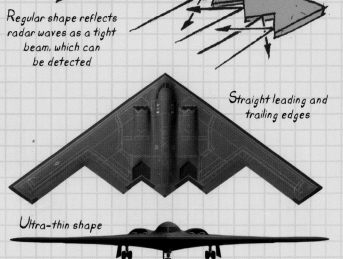

Ultra-thin shape

AIRBUS A380

The world's largest passenger plane, the A380 Super Jumbo, first flew in 2005 and went into regular service in 2007. It's also known as a "double-decker," with two passenger cabins almost the full length of its fuselage. In fact, it's a triple-decker, with all the luggage on the lowermost third deck—the cargo compartments under the passenger floor.

Eureka!

In 1940, the Boeing 307 Stratoliner was the very first airliner to have a pressurized cabin. This meant it could fly above thunderstorms without passengers suffering from lack of oxygen.

What next?

Airbus is planning a "stretched" version of the A380, which will be 21 feet (6.4 m) longer, with room for 150 more passengers.

Passenger cabin There's room for 555 people in the usual three classes of seats, or 853 if they all travel in economy class seats.

Flight deck The seats, controls, and instruments for the two pilots are similar to other Airbus planes such as the A340 and A320. This allows pilots to swap aircraft more easily.

At takeoff a fully loaded A380 weighs 600 tons and needs a runway almost 2 mi long.

Wheel fairing

☀ FLY-BY-WIRE

In some planes the pilot's control column (joystick), rudder pedals, wheel brake lever, and other controls are connected by long metal cables to the parts they move, such as the elevators, ailerons, and rudder. In a fly-by-wire aircraft, the pilot's controls send signals along electrical wires into a computer. This then sends out signals to electric motors and hydraulic pumps, which work the moving parts. The computer helps by alerting the pilots to problems.

The A380's fly-by-wire flight deck has eight screen displays

When the Boeing 747 first flew in 1969, it became known as the Jumbo Jet. The A380 is even bigger, so it's nicknamed the Super Jumbo.

Take a tour of the incredible flight deck of the A380 by visiting www.factsforprojects.com and clicking the web link.

Seats Even the cheapest economy seats have their own 10-inch (25.4 cm) flat screen for movies and TV, computer USB connections, and power sockets to charge gadgets such as MP3 players.

Fin 79 ft (24 m) tall

Construction Many parts of the airframe are composite materials such as carbon fiber, for less weight combined with great strength.

The A380 can seat 538 people on the lower deck and 315 on the upper (if everyone sits in economy class).

Flexible wings In high winds, the wings can bend up and down safely by more than 3 feet (0.9 m). The wingspan is 262 feet (80 m)—36 feet (11 m) more than the latest Boeing 747 Jumbo Jet.

The Airbus A380 began service flying between Singapore and Sydney. First tickets sold on eBay for up to $100,000.

The A380 has 22 wheels, four more than the 747.

Engine pylon

Fan shroud

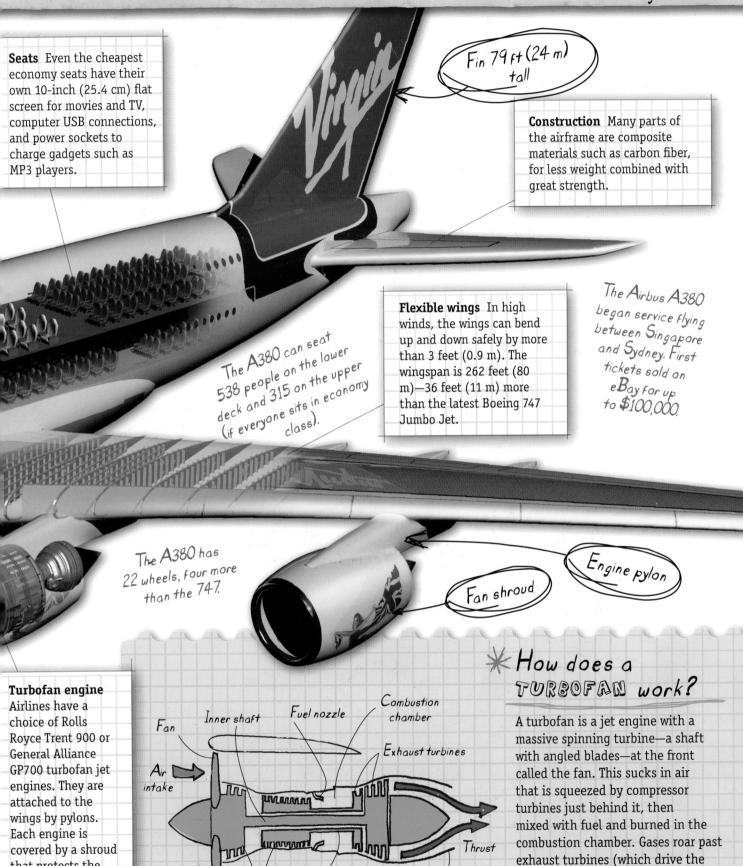

Turbofan engine Airlines have a choice of Rolls Royce Trent 900 or General Alliance GP700 turbofan jet engines. They are attached to the wings by pylons. Each engine is covered by a shroud that protects the front fan and forms the outer casing.

Fan
Inner shaft
Fuel nozzle
Combustion chamber
Air intake
Exhaust turbines
Thrust
Outer shaft
Exhaust nozzle
Fan shroud
Compressor turbines

✳ How does a TURBOFAN work?

A turbofan is a jet engine with a massive spinning turbine—a shaft with angled blades—at the front called the fan. This sucks in air that is squeezed by compressor turbines just behind it, then mixed with fuel and burned in the combustion chamber. Gases roar past exhaust turbines (which drive the compressors) and out of the back to push the engine forward. The huge front fan also works like a propeller for added thrust.

F-35B LIGHTNING

One of the world's most advanced aircraft, the F-35 Lightning fighter-bomber first roared into the skies in 2006. The F-35B is a special STOVL version—Short Takeoff and Vertical Landing. Using the lift fan just behind the pilot, it needs only a very short runway to become airborne, and it is able to land straight down.

Eureka!

The idea of a lift fan for vertical takeoff and landing was first tried in the 1920s with propeller planes. However, the piston engines of the time could not produce enough power to drive both the propellers and the fan.

What next?

The F-35 Lightning shares "stealth" features with the F-22 Raptor and even newer, secret "X-planes" still at the design stage.

Twin fins Like many modern combat aircraft, the Lightning has two fins, each with a rudder. This gives better control, especially at low speeds when taking off and landing on an aircraft carrier.

All-moving tailplane The whole tailplane (small wing at the back) tilts to work as an elevator for super-quick climbs, dives, and maneuvers.

Rudder

Wheel

✳ How does a LIFT FAN work?

The F-35B's lift fan looks like a big cooling fan with blades set at an angle. It is driven by a spinning shaft from the main engine. Air is sucked in from above and forced out at high speeds below to push the plane upward. The clutch connects it to the engine for use, then disconnects it for normal flight.

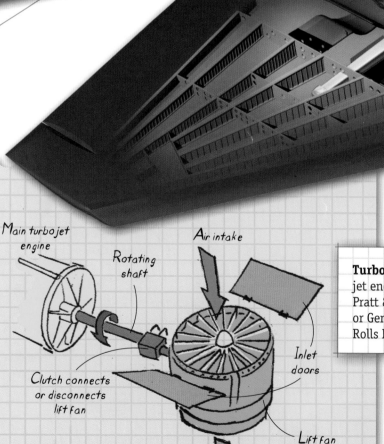

Main turbojet engine

Rotating shaft

Air intake

Clutch connects or disconnects lift fan

Inlet doors

Lift fan

Air outlet produces thrust

Turbojet The single jet engine is either a Pratt & Whitney F135 or General Electric-Rolls Royce F136.

The F-35B (the version with the lift fan) first flew in June 2008, 18 months after the F-35.

Watch a video of the Lightning in action by visiting www.factsforprojects.com and clicking the web link.

Sensors Small sensors with cameras, lasers, and heat detectors are scattered all over the F-35B. They pick up any objects moving nearby, from friendly planes to enemy missiles.

The name "Lightning" has been used for previous war planes including the prop Lockheed P-38 Lightning of World War II and the English Electric Lightning jet of the 1960s.

✳ How does a TURBOJET work?

A turbojet is similar to a turbofan (see page 33). Air is sucked in at the front, gets squeezed or compressed, burns with fuel, and roars out the back to produce thrust. Turbojets lack a big fan at the front, so they can go much faster than turbofans. However, they are noisier and consume more fuel.

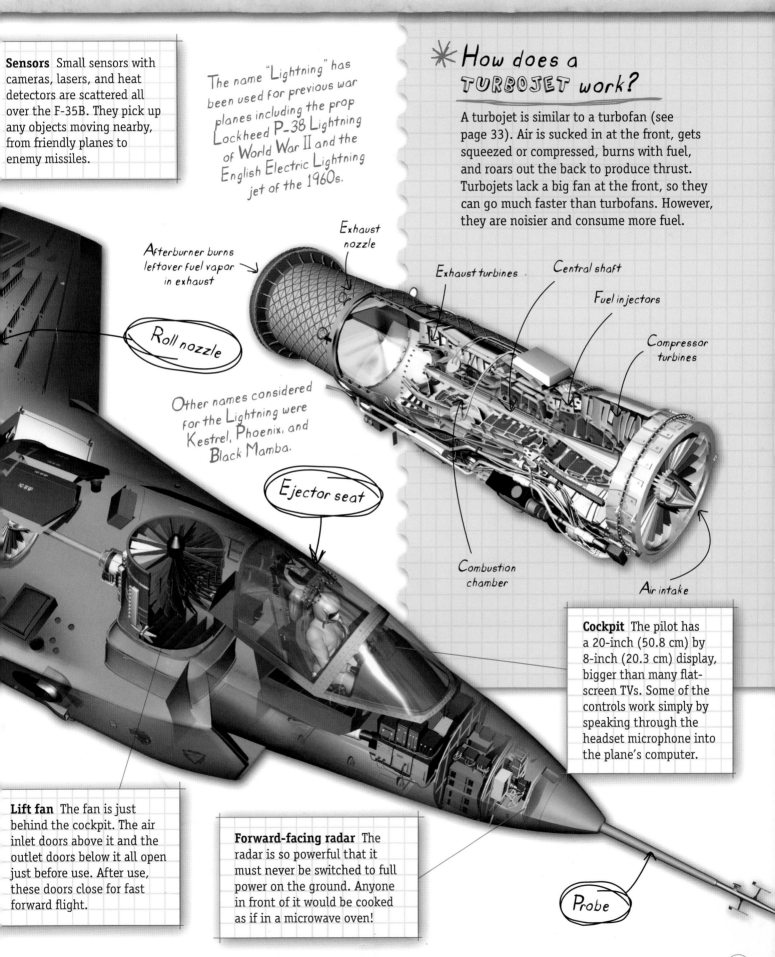

Exhaust nozzle

Afterburner burns leftover fuel vapor in exhaust

Roll nozzle

Exhaust turbines

Central shaft

Fuel injectors

Compressor turbines

Other names considered for the Lightning were Kestrel, Phoenix, and Black Mamba.

Ejector seat

Combustion chamber

Air intake

Cockpit The pilot has a 20-inch (50.8 cm) by 8-inch (20.3 cm) display, bigger than many flat-screen TVs. Some of the controls work simply by speaking through the headset microphone into the plane's computer.

Lift fan The fan is just behind the cockpit. The air inlet doors above it and the outlet doors below it all open just before use. After use, these doors close for fast forward flight.

Forward-facing radar The radar is so powerful that it must never be switched to full power on the ground. Anyone in front of it would be cooked as if in a microwave oven!

Probe

BOEING CH-47 CHINOOK

The Chinook heavy-lift helicopter is used by air forces and civilians in more than 20 countries worldwide. A helicopter's spinning rotor blades are shaped like long, thin plane wings. They provide lift in the same way as they spin around. This is why helicopters are called rotary-wing aircraft.

Eureka!

In 1939, Russian aircraft pioneer Igor Sikorsky came up with the basic helicopter design still used today. He also designed successful flying boats.

If one engine fails, the Chinook can still fly with the other one driving both rotors.

Rotor The rotor's three fiberglass blades make a circle 60 feet across and spin almost four times each second. Their airfoil shape produces lift as they turn, and their chop-chop-chop sound is why helicopters are called "choppers."

Front rotor head The front rotor spins one way, which makes the helicopter body spin the other way. The rear rotor spins in the opposite direction, and the two cancel each other out. This means there is no need for a small upright rotor as on other helicopters.

Apart from military missions, Chinooks carry people and supplies in disaster areas and are also used for logging, firefighting, and surveys.

The rotor in an autogyro relies on passing air for lift

Nose gun

Machine gun hatch

Flight deck The Chinook usually has a crew of three. The main pilot sits in the left seat, with the copilot in the right. The flight engineer is in the compartment behind them.

✴ What is an AUTOGYRO?

Like a helicopter, an autogyro has a rotor, but this is not engine-powered—only the propeller is. As the propeller pushes the autogyro forward, passing air makes the rotor whirl around for lift.

What next?

The Solotrek strap-on personal helicopter has two rotors in ring-shaped casings above the pilot's head, fixed onto a chairlike frame. But it's not available in a store near you just yet!

Make a model helicopter by visiting www.factsforprojects.com and clicking on the web link.

The Chinook is named after winds in northwest North America. They are warm, dry, and gusting, and blow downward like the craft's rotor blast.

Cabin Almost 85 feet (26 m) long, the straight-through, open cabin can take more than 50 fully equipped troops, several vehicles, or 13 tons (11.8 metric tons) of cargo. Top speed is more than 180 mph (289.7 km/h).

Engines There are two Lycoming turboshaft engines, one on each side of the tail fin base. They have turbines, like jet engines. They turn an axle-like shaft rather than use their hot gas blast for thrust.

Gearbox The gearing system changes the very fast revolutions of the engine to the much slower, more powerful turning force for the rotors.

629089

Rear loading ramp

UNITED STATES ARMY

Missile pod

✳ How does the ROTOR HEAD work?

Connecting rods link the turning rotor blades and their ring-shaped upper swashplate to the non-turning lower plate below. As the pilot's controls lift, lower, or tilt the whole swashplate, the blades also tilt, altering their amounts of lift. This makes the helicopter rise, descend, hover, and go in any direction.

Blade

Rotor head

Connecting rod

Blade tilts

Upper swashplate

Nonrotating lower swashplate

Direction of blade rotation

Spinning rotor shaft

GLOSSARY

Airfoil

The shape of most aircraft wings, being more curved or humped on the upper surface than the lower surface, to provide a lifting force.

Aileron

The control surface of an aircraft, usually on the trailing (rear) edge of the wing, that makes it lean left or right (roll or bank).

Airframe

The strong inner framework or "skeleton" of an aircraft.

Battens

Long, thin strips of wood, plastic, or a similar material that help to hold out a flexible surface, such as a hang glider wing or yacht sail.

Biplane

An aircraft with two sets of main wings, usually one above the other.

Blade

One of the long, slim parts of a propeller (airscrew) or helicopter rotor; some propellers have six blades or more.

Bracing wires

Thin metal wires or cables that are stretched tightly between various parts of an aircraft, especially the wings, to hold them steady.

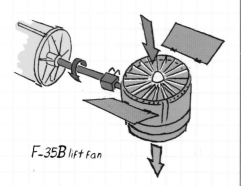

F-35B lift fan

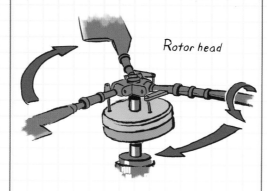

Rotor head

Canard design

An aircraft layout with the small wings or tailplane in front of the large main wings rather than behind them.

Chord

On an aircraft, the measurement from the front or leading edge of the wing to its rear or trailing edge.

Cockpit

The control compartment of an aircraft where the pilot sits. On larger aircraft with two or more flying crew, it is sometimes called the flight deck or control cabin.

Cocoon

A long bag, similar to a sleeping bag, for the pilot of a hang glider or similar craft.

Drag

The force or resistance of air pushing against something moving through it.

Elevator

The control surface of an aircraft, usually on the tailplane (small rear wing) that makes it tilt up or down (pitch).

Elevon

The control surface of a tail-less or flying-wing aircraft, which is a combined elevator and aileron.

Envelope

In ballooning, the main casing of the balloon that contains the hot air or lighter-than-air gas.

Fin

The upright part at the rear of most aircraft, also known as the vertical stabilizer and often called the tail.

Fuselage

The main body or central part of an aircraft, usually long and tube shaped.

Gores

Curved strips that make up the envelope (main part) of a balloon.

Hub

The central part (boss) of a propeller, wheel, or similar spinning object, where it turns on its shaft or axle.

Infrared

A form of energy, as rays or waves, which is similar to light but with longer waves that have a warming or heating effect.

Monoplane

An aircraft with one main pair of wings, rather than two (biplane), three (triplane), or more.

Radial piston engine

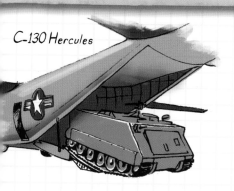

C-130 Hercules

Pitch

When an aircraft points up or down, to climb or descend. Also the angle of a propeller blade compared to the prop's direction of movement.

Pusher prop

A propeller or airscrew at the rear of an engine that pushes air backward, rather than one at the front that pulls its way through the air.

Radar

A system of sending out radio waves that reflect off objects and detect the echoes to find out what is within a particular area.

Retractable

When something folds away or retracts into a compartment, such as aircraft wheels, to leave a smooth surface.

Rib

The short, curved parts inside a wing, from the front or leading edge to the rear or trailing edge.

Roll

When an aircraft leans or banks to one side.

Rudder

The control surface of an aircraft, usually on the upright fin or "tail," which makes it steer left or right (yaw).

Spar

The long, rigid, beam-like part inside a wing, from root to tip, that gives it strength.

Spinner

The cone-shaped covering over a propeller's central part, the hub (boss), to protect it and improve streamlining.

Stealth technology

Designing an aircraft, ship, or similar object so it is difficult to detect by sight, sound, heat sensors, or radar equipment.

STOL

Short Takeoff and Landing, when an aircraft can take off and land in a very small distance.

Tailplane

The two small rear wings on most aircraft, also known as the horizontal stabilizers. They carry the elevators and are usually next to the fin or "tail."

Thrust

A force that pushes an object forward, such as the propellers or jet engines of an aircraft.

Turbofan

A jet engine with fanlike turbine blades inside and one very large turbine or "fan" at the front that works partly as a propeller.

Adjustable pitch prop

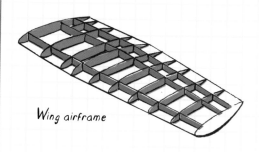

Wing airframe

Turbojet

A jet engine with fanlike turbine blades inside, which produces a powerful blast of gases from the rear end.

Turboprop

A jet engine with fanlike turbine blades inside, which turns a propeller for thrust rather than using its jet blast of gases.

Turboshaft

A jet engine with fanlike turbine blades inside, which spins a shaft for power rather than using its jet blast of gases.

Undercarriage

An aircraft's landing wheels, skids, floats, or similar devices that support it on the ground.

VTOL

Vertical Takeoff and Landing, lifting off and touching down by going vertically, straight up and down.

Winglet

A small angled-up or "bent" part at the tip of a wing.

Wingspan

The distance of an aircraft's main wings from one tip to the other.

Yaw

When an aircraft steers to the left or right (like a car).

INDEX